Status and Needs of Forensic Science Service Providers: A Report to Congress

Contents

Executive Summary 2

Introduction 9

Manpower and Equipment Needs 12

Continuing Education 17

Professionalism and Accreditation Standards 24

Collaboration Among Federal, State, and Local Forensic Service Providers 28

Summary of Forensic Community Recommendations 30

Executive Summary

Introduction

The 2004 Consolidated Appropriations Act, H.R. 2673 requires the National Institute of Justice (NIJ) to transmit to Congress a report addressing the needs of forensic service providers beyond the DNA initiative (see *Advancing Justice Through DNA Technology,* available on www.dna.gov or www.usdoj.gov/ag/dnapolicybook_cov.htm). Specifically, the Act states:

> Improving Forensic Capabilities - The National Institute of Justice [NIJ], in conjunction with its own Office of Science & Technology, the American Society of Crime Lab Directors, the American Academy of Forensic Sciences, the International Association for Identification, and the National Association of Medical Examiners, is directed to develop a plan which will address the needs of the crime lab and medical examiner community beyond the "DNA Initiative" and report back to the Committees on Appropriations no later than 180 days from the date of enactment of this Act. The report should address the following: (1) manpower and equipment needs, (2) continuing education policies, (3) professionalism and accreditation standards, and (4) the level of collaboration needed between Federal forensic science labs and State/local forensic science labs for the administration of justice.

Over the course of four months, NIJ collaborated with each of the organizations named by Congress, including the American Academy of Forensic Sciences (AAFS), the American Association of Crime Laboratory Directors (ASCLD), the International Association for Identification (IAI), and the National Association of Medical Examiners (NAME). At a summit held in Washington, D.C., on May 18–19, 2004, each organization presented their formal comments. The summit also included input from the Bureau of Justice Statistics (BJS) concerning the data contained in their reports, *50 Largest Crime Labs 2002 and Census of Publicly Funded Forensic Crime Laboratories,* which surveyed 350 labs and is not yet published.

The findings in this report represent the opinions of the forensic community presented at the summit, not necessarily the views of the Department of Justice or the National Institute of Justice. The summit reports, presentations, agenda, and participant list can be found on the National Institute of Justice Web site at www.ojp.usdoj.gov/nij.

This study covers a wide range of forensic disciplines, including general toxicology, firearms/toolmarks, questioned documents, trace evidence, controlled substances, biological/serological screening, fire debris/arson analysis, impression evidence (e.g., fingerprints, shoe/tire prints), blood pattern analysis, crime scene investigation, medicolegal death investigation, and digital evidence. It should be noted that not all forensic services are performed in what is thought of as a traditional crime laboratory. Forensic services in the disciplines of digital evidence, latent prints, questioned documents, and crime scene investigation may also be provided at a site outside of the traditional crime laboratory setting by a unit composed of sworn law enforcement personnel who may or may not have scientific training.

Thus, this report may not represent a complete view of the needs and challenges of those particular forensic disciplines. In the critical area of digital evidence, NIJ published the *Electronic Crime Needs Assessment for State and Local Law Enforcement* in 2001 and established the Electronic Crime Partnership Initiative to identify and track these needs on an ongoing basis. This document is available online at http://www.ojp.usdoj.gov/nij/pubs-sum/186276.htm.

Further, all summit participants agreed that there was insufficient time to develop quantifiable data. The conclusions are largely the opinions of local, state and federal forensic science practitioners who participated directly in the study, Thus, it is not possible to present a full and complete picture in certain areas. For example, the AAFS noted that questioned documents and firearms examiners responded well to their questionnaire, but only one toxicologist returned a completed survey. It is estimated that there are at least 1,000 forensic service providers in the United States. The forensic science organizations recommend that a concerted effort be made to identify all forensic service providers in order to raise the level of awareness of the quality and quantity of services being provided to the criminal justice community.

Forensic Community Recommendations

The forensic sciences enjoy great visibility and respect among the public today. Popular television shows depict the crime laboratory as an important and exciting endeavor, and young people are choosing to study forensic science in college in unprecedented numbers. In particular, DNA analysis has revolutionized the ability of law enforcement to identify criminals and protect the innocent from wrongful prosecution.

Nonetheless, crime laboratories face several important challenges. First and foremost, the forensic service organizations identified personnel needs, as well as education and training for new forensic scientists, as long-standing problems. Although it is difficult to quantify these needs, every forensic discipline believes that it faces shortfalls of personnel qualified to replace retiring examiners or meet increasing case workloads. In addition, examiners should be required to meet minimum training and proficiency standards in all disciplines. The 1999 NIJ report, *Forensic Sciences: Review and Status of Needs,* contained recommendations concerning training needs that are still valid, according to ASCLD.

Also, the forensic service organizations recognize the need to improve the scientific understanding of the scientific foundations of specific disciplines. DNA analysis has a fully characterized statistical and scientific basis, in that the uniqueness of one individual's DNA profile can be quantified and presented with great accuracy. Scientific research and the publication of best practices guides can improve the practice and acceptance of the forensic disciplines.

These issues should be addressed more fully by the Forensic Science Commission which is authorized by the Justice For All Act. Each of the forensic service organizations supports the creation of a Forensic Science Commission to review the needs of the forensic science community in the long term at the federal, state and local levels. As part of the DNA Initiative, the President has called for the creation of a Forensic Science Commission. The Commission will be charged with two primary responsibilities: (1) developing recommendations for long-term strategies to maximize the use of current forensic technologies to solve crimes and protect the public, and (2) identifying potential scientific breakthroughs that may be used to assist law enforcement. The Commission is viewed by the organizations as a mechanism to identify issues and needs of particular disciplines and provide national leadership to improve the practice of forensic science.

Manpower and Equipment

Manpower shortages are the biggest concern of the forensic community and directly impact on the ability of crime laboratories to address casework backlogs. According to preliminary results from the upcoming BJS report on the 50 largest crime labs, by the end of 2002, crime laboratories reported a backlog of about 270,000 forensic analysis requests. (For the purpose of the laboratory census, a backlog was defined as any request that remained unanalyzed in the laboratory for more than 30 days.) The laboratories, which employed 4,300 full-time equivalent personnel, reported that they would need approximately 930 additional full-time equivalents (at an estimated cost of approximately $36 million), to achieve a 30-day turnaround for 2002 requests. All member organizations reported equipment shortages as a limiting factor in processing forensic casework. Specifically, ASCLD estimates that equipment needs for the 50 largest crime laboratories in the disciplines of controlled substances, trace evidence, firearms, questioned documents, latent prints, toxicology and arson exceed $18 million. ASCLD also recommended that a reliable process be established to monitor the manpower and equipment needs of the forensic community on an ongoing basis.

The forensic community reports even more acute manpower shortages for the death investigation system. NAME reports that the United States requires at least 850 board-certified forensic pathologists, roughly double the current number. Many autopsies are now performed by individuals without needed training in general pathology and forensic pathology. Equipment is lacking in some basic areas of need, such as histology, microbiology, clinical lab testing, and genetic and metabolic services.

In general, the forensic science community is concerned about improving its capacity, an issue that relates to manpower and equipment, as well as other issues covered in this study. The organizations support the continuation or expansion of Coverdell funding to support specific needs, including: fingerprint identification systems, alternate light sources, vehicles, training, accreditation and certification, and photo and digital imaging equipment. The value of DNA as a forensic tool is also recognized and supported as part of an overall funding strategy. Other forensic community recommendations include:

- Crime laboratories need dedicated staff to support quality assurance and accreditation programs.

- Certain forensic disciplines appear to have important manpower shortfalls, including crime scene processing, digital evidence analysis, latent fingerprint examination, firearms examination, document analysis, and toxicology.

- The FBI should increase the number of Universal Latent Workstations to state and local law enforcement so that the full capacity of the International Automated Fingerprint Identification System (IAFIS) may be utilized.

- The federal government should work to ensure interoperability among automated fingerprint identification systems (AFIS) of different manufacture, as well as the interoperability of these systems with IAFIS.

- Through the Forensic Science Commission or another means, forensic science providers outside crime laboratories should be identified and advised of professional and governmental assistance programs.

Continuing Education

The forensic community reports that training needs are significant across all disciplines. This includes training of novices and continuing education for experienced professionals. In particular disciplines, such as questioned documents, there are a declining number of qualified experts, according to AAFS.

The Technical Working Group on Education and Training (TWGED) recommended that between 1 percent and 3 percent of the total forensic science laboratory budget be allocated for training and continuing professional development. Preliminary data reported by BJS from its crime laboratory census showed that the training and continuing education budgets of the largest 50 laboratories in the United States were actually less than one-half of 1 percent of their total budgets. To close this gap, according to the forensic science organizations, the federal government should provide grants for continuing education or training academies for the forensic sciences. For example, AAFS indicated a need for a central and/or west coast training academy, similar to the current east coast academy under the Bureau of Alcohol, Tobacco, and Firearms. Regional or technology-based academies should be established, using the FBI Academy as a model.

Some options to address the training needs of forensic examiners and managers include traditional face-to-face or hands-on training, collaborations, and alternative delivery systems such as electronic media. Regional centers would be suited for expanding the scope and delivery of training programs. Also, professional models for training and establishing competency should be encouraged. The forensic science community should consider methods to encourage quality graduate education in forensic science. ASCLD suggested that a program to eliminate or forgive student loans for graduates who obtain full-time employment in public forensic science laboratories be considered. Other forensic community recommendations include:

- Minimum standards should be established for each forensic discipline for equipment, techniques, training and documentation. These standards should include testing of personnel to confirm minimum competency. In particular, NIJ and FBI should collaborate with their scientific working groups to generate and implement standards throughout the forensic sciences.

- The FBI should increase the number of Universal Latent Workstation systems to state and local law enforcement so that the full capacity of the International Automated Fingerprint Identification System (IAFIS) may be utilized.

- Forensic science training programs at the FBI should be "reactivated" (IAI) or expanded.

- Tuition assistance should be provided to encourage enrollment in university forensic science degree programs.

Professionalism and Accreditation Standards

Each of the forensic science organizations supports the exploration of mandatory accreditation of organizations and certification of practitioners. Accreditation is a voluntary program through which a laboratory demonstrates that its management, operations, personnel, procedures, equipment, physical plant, security, and health and safety procedures meet established standards. Certification is a process of peer review through which an individual practitioner is recognized as having attained the professional qualifications needed to practice in one or more disciplines.

The organizations also support funding to support quality assurance programs to help labs attain accreditation. Maintaining and increasing professionalism within the forensic science community is critical to the delivery of quality services. Professionalism is enhanced by demonstrating compliance with quality assurance measures such as laboratory accreditation and practitioner certification. Unfortunately, many laboratories are confronted with budgets that are insufficient to meet caseload demands and at the same time support participation in accreditation and certification programs. Costs associated with accreditation and certification programs include proficiency testing and inspection fees, at a minimum. Dedicated personnel are needed to support participation in such programs, and examiners need to be given the time away from casework to participate in proficiency testing programs.

The Forensic Science Commission can identify strategies to address these needs in coordination with ongoing activities, especially those in NIJ and the FBI.

Collaboration Among Federal, State, and Local Forensic Service Providers

Federal laboratories collaborate with State and local forensic service providers in many ways. They provide leadership and resources for research, training, and technology transfer. Federal laboratories also maintain and support investigative databases for firearms, fingerprints, and DNA. The FBI has provided onsite training and online training via its Virtual Academy. Over the years, the forensic science organizations maintain that the FBI has decreased training available to State and local agencies. The forensic community would like the Federal forensic science training programs expanded to meet current and future needs. Specifically, they recommend:

- The federal government must strengthen the support given to crime labs and other crime scene/disaster scene first responders with respect to terrorism or other events that might result in mass casualties, including support for training, equipment and coordination activities. Of particular concern is the training of crime scene responders in the safe handling of evidence that may be contaminated with biological, chemical, or radiological material.

- Forensic science providers need greater awareness of state and federal assistance and programs, especially those outside the traditional crime laboratory.

- The federal government should conduct scientific research to improve the practice of forensic science and address emerging technology challenges from criminals, particularly in the area of electronic crime. The federal government should also play a leading role in advocating interoperability and information sharing, such as in automated fingerprint identification systems.

Research and Development

Although Congress did not specifically ask for input concerning research and development needs, each of the forensic science organizations outlined specific needs for improved scientific understanding and technology to serve the forensic community. In particular, the following needs were outlined:

- Basic research is needed into the scientific underpinning of impression evidence, (especially fingerprint evidence, but also footwear and tire track evidence), questioned documents, and firearms/toolmark examination.

- NIJ should continue its program to develop a fast live scan device to collect forensic-quality fingerprints and palm prints. NIJ is currently soliciting research and development proposals for this technology.

- The federal government should sponsor research to validate forensic science disciplines to address basic principles, error rates, and standards of procedure.

- Crime laboratories need tools to improve speed and efficiency, extend forensic analysis to more difficult samples, and support the full range of forensic techniques. Technology is needed to improve evidence collection, crime scene analysis, field testing of drugs and other material for investigative purposes.

Summary

The forensic sciences community believes that it needs additional attention. The Forensic Science Commission should help provide national leadership by monitoring the needs of the forensic science community on an ongoing basis. The Commission will bring together providers and end users of forensic services to address issues raised in this report that affect the quality and timeliness of oar Nation's forensic services. The Commission should consider issues that affect all disciplines of the forensic sciences and make recommendations to improve public safety through maximizing the use of forensic evidence. Its multidisciplinary membership will facilitate the development of strategic partnerships that represent diverse opinions and perspectives, including those of law enforcement, practitioners, academicians, attorneys, judges and ethicists. Such partnerships will be able to contemplate in a public forum complex issues that affect the furtherance and advancement of forensic practice. It is recognized that a number of social, ethical, legal, and policy issues may arise as a result of effectively and efficiently implementing recommendations for enhancing our Nation's forensic service providers. However, NIJ and the forensic organizations are confident that the Forensic Commission will serve to provide a cognitive process by which criminal justice professionals and the public can openly and fully deliberate these critical matters.

Introduction

Forensic disciplines discussed at the summit held at NIJ in collaboration with 12 representatives of the forensic science community include general toxicology, firearms/toolmarks, questioned documents, trace evidence, controlled substances, biological/serology screening, fire debris/arson analysis, impression evidence (e.g., fingerprints shoe/tire prints), blood pattern analysis, crime scene investigation, medicolegal death investigation, and digital evidence. Each of the member organizations was responsible for developing data and information covering several disciplines. AAFS addressed general toxicology, firearms/toolmarks, and questioned documents. ASCLD addressed trace evidence, controlled substance, biological/serological screening, and fire debris/arson analysis. IAI addressed impression evidence, blood pattern analysis, crime scene investigation, and digital evidence. NAME reported on the issues affecting the medical examiner and coroner communities.

Participants in the study

AAFS is a professional society dedicated to the application of science to the law and is committed to the promotion of education and the elevation of accuracy, precision, and specificity in the forensic sciences. Founded in 1948, AAFS has a membership of more than 5,500 forensic sciences professionals located principally in the United States with members in 56 other countries. AAFS comprises 10 sections representing a range of forensic specialties. The members are physicians, attorneys, dentists, toxicologists, physical anthropologists, document examiners, psychiatrists, engineers, criminalists, educators, and others who practice, study, and perform research in the forensic sciences.

ASCLD is a nonprofit professional society dedicated to providing excellence in forensic science analysis through leadership in the management of forensic science. Now in its 32nd year, ASCLD has 550 members representing 245 local, State, Federal, and private crime laboratories in the United States. Membership includes directors from 30 international laboratories and national and international academic affiliates. The purpose of the organization is to foster professional interests; assist the development of laboratory management principles and techniques; acquire, preserve, and disseminate forensic-based information; maintain and improve communications among crime laboratory directors; and promote, encourage, and maintain the highest standards of practice in the field.

IAI membership comprises more than 5,600 individuals from 70 nations and 13 forensic disciplines. IAI offers training and educational opportunities in fingerprints, crime scene investigation, forensic photography and electronic imagining, firearms and toolmarks, bloodstain pattern identification, footwear and tire track analysis, questioned documents, polygraph, forensic art, forensic odontology, innovative and general techniques, and laboratory analysis.

NAME is the primary professional organization for medical examiners and forensic pathologists in the United States. Established in 1965, NAME currently has nearly 1,000 members, of which approximately 80 percent are physicians and 20 percent are affiliated lay death investigators or administrators who work in medical examiners' offices. The medical examiner community is a unique group of professionals within forensic service providers, and their issues are discussed separately in some chapters. The daily practice of forensic pathology extends far beyond questions related to medicine and forensic pathology and often involves dealing with political entities, the media, law enforcement, the judicial system, health care systems, families of the deceased, and members of the general public.

Forensic Disciplines

General toxicology. Toxicology involves the examination of body fluids or tissues for the presence and quantity of substances such as drugs or poisons in ante- or postmortem casework. Examples include body fluids such as blood, urine, and spinal fluid, and organ and muscle tissue.

Firearms/toolmarks. Firearms identification determines whether an evidence bullet was fired from a suspect weapon. It may also include comparison of fired cartridge cases, firearm function tests, serial number restorations, and distance determinations. Toolmarks left at a crime scene or on a victim by various types of implements (e.g., knives, screwdrivers, pliers) can be microscopically compared to test marks made in the laboratory by suspect tools. The forensic scientist is then able to determine if a suspect tool was used in the commission of a crime.

Questioned documents. A questioned document contains a signature, handwriting, typewriting, or other mark whose source or authenticity is in dispute or doubtful. The forensic document examiner makes examinations, comparisons, and analyses of documents to establish genuineness, expose forgery, or reveal alterations. Letters, checks, driver licenses, contracts, wills, voter registrations, passports, petitions, threatening letters, suicide notes, and lottery tickets are common types of questioned documents.

Trace evidence. Trace evidence is physical evidence that results from the physical transfer of small or minute quantities of materials (e.g., hair, textile fibers, paint chips, glass fragments). This category of evidence encompasses many diverse types of microscopic materials as well as some examples that are easily visible to the naked eye.

Controlled substances. In the discipline of controlled substance identification, evidence is examined to identify drugs, either prescription drugs such as Valium or illegal drugs such as cocaine. Evidence examples might include plant material, powder, drug paraphernalia, tablets, and pills.

Biological/serology screening. This discipline encompasses a variety of tests to determine the presence of blood, semen, saliva, or other body fluids. Chemical and microscopic methods of testing are often used to determine whether samples are suitable for subsequent DNA testing.

Fire debris/arson analysis. Arson analyses include the examination and testing of items and debris collected from a fire scene. The scientist tests materials to determine if an ignitable material is or was present, which can help investigators determine whether a fire was deliberately set.

Impression evidence. Impression evidence involves objects or materials that have retained the characteristics of other objects that have been physically pressed against them (e.g., fingerprints, shoe/tire prints). A latent print is an impression that is not readily visible, made by contact of bare hands or feet with a surface resulting in the transfer of materials from the skin to that surface. Footwear or tire track impressions from a crime scene can be found on many types of material, such as hard flooring, dirt, mud, and dust.

Blood pattern analysis. Blood pattern is the analysis of stains left by blood shed at a crime scene. Bloodstain patterns can yield valuable information for the reconstruction of the incident. Bloodstain pattern analysis may clearly define the location of the victim or the assailant(s) by establishing their actions.

Crime scene investigation. Crime scene investigation involves the recovery and analysis of forensic evidence, in addition to addressing issues such as security, prevention of contamination, locating and collecting items of evidence, interpretation of evidence, and possible reconstruction of the event. Crime scene investigation provides the best opportunity to determine actual events associated with the commission of a crime.

Medicolegal death investigation. The medical examiner community is a unique group of professionals who play an important role in the investigation of sudden, unnatural, unexplained, or suspicious deaths, including homicides, suicides, unintentional injuries, drug-related deaths, and other deaths that are sudden or unexpected by determining the cause and manner of death. In many jurisdictions, responsibility for conducting death investigations may rest with pathologists, medical examiners, or coroners.

Digital evidence. The discipline of digital evidence includes all facets of crime where evidence may be found in a digital form. It includes forensic imaging, forensic audio and video analysis, and analyzing computer files and other digital data from computer systems.

1. Manpower and Equipment Needs

Crime laboratories need equipment and manpower to perform their work. Across all disciplines of forensic science, having an adequate number of personnel and specialized equipment ensures that the best forensic services are provided to the criminal justice community. Personnel should be adequately educated and trained. Equipment should be affordable and reliable.

When the demand for service exceeds a service provider's capacity to analyze the submitted evidence, backlogs result. These "casework" backlogs cause significant delays in the courts as well as in the investigation of crimes. To address backlog issues, crime laboratories often employ a variety of strategies to manage demand. For example, laboratory work is often prioritized according to court dates. Many laboratories establish case acceptance policies to limit the number of cases coming into the laboratory. As a result, in many cases where no suspect has been identified, evidence is not even brought into the laboratory by police agencies. In extreme cases, the laboratory may sometimes return evidence if it cannot be analyzed in a timely manner.

The forensic community has identified personnel as its primary need. According to preliminary results from an upcoming Bureau of Justice Statistics (BJS) survey of the largest 50 crime laboratories in the United States, *50 Largest Crime Labs 2002,* the laboratories received more than 994,000 new cases in 2002, including more than 1.2 million requests for forensic services. (A case is evidence from a criminal investigation; a "request" is a request for a specific type of analysis within that case, e.g., controlled substances, latent prints.) The survey information will be part of the larger BJS *Census of Publicly Funded Forensic Crime Laboratories,* which surveyed 350 labs and will be published in the future.

The Bureau of Justice Statistics reported that laboratories began 2002 with 115,000 backlogged requests for forensic service, received an additional 1.2 million requests, and completed 1.1 million requests, ending 2002 with a backlog of about 270,000 requests. (For the purpose of the BJS crime laboratory census, a backlog was defined as any request that remained unanalyzed in the laboratory for more than 30 days). About 80 percent of the estimated 270,000 backlogged requests for forensic service were attributable to controlled substances (50 percent), latent prints (18 percent), and DNA analysis (11 percent). Backlogs were also seen in firearms/toolmarks, toxicology, and trace. Other needs identified include additional equipment ($18 million), supplies, laboratory space, overtime, travel, and training. The crime labs, which employed 4,300 full-time equivalent personnel, estimated that about 930 additional full-time equivalents would be needed to achieve a 30-day turnaround for 2002 requests. Estimated cost for the additional personnel would be approximately $36 million.

According to an IAI survey of its members, a large number of latent fingerprint cases are backlogged, with the largest backlogs in the largest agencies. In the largest 12 organizations, backlogs range from several hundred to 1,000 cases. The average backlog time in these agencies is 166 days with total backlogs of 5,147. Agencies do their best to prioritize serious crimes against persons before property crimes, but that often is not effective. Six of these large organizations are service centers for a number of law enforcement agencies, so their backlog reflects back on their customer agencies. Of particular concern is the lack of IAI-certified fingerprint examiners to fill many vacant positions in the United States. Agencies are increasingly seeking certified latent examiners.

Common Concerns Across All Disciplines

In forensic science, cost-effectiveness and budgetary constraints are constant concerns. Although each of the forensic disciplines concentrates on different evidence types and has specific personnel, training, equipment, and facility requirements, they share common concerns about continuing education, adequate manpower, equipment, and training. For example, manpower and equipment issues discussed during the summit include the following:

- The forensic community believes that the number of nonlaboratory forensic service providers needs to be determined. These providers perform many analyses in criminal matters and need to be identified. An IAI survey of its members found that approximately 66 percent of fingerprint identification is not done in a traditional crime laboratory setting, but rather in police and sheriff's departments and State crime bureaus. Such fingerprint work is performed by units with titles such as Crime Scene Unit, Identification Division or Unit, or Fingerprint Unit. Many fingerprint examiners in these types of units are sworn law enforcement officers. According to the IAI survey, the average staff size of such a unit is 9.1 with the largest unit at 51 and the smallest unit at 1. Many of the personnel in these units testify in court to latent fingerprint identifications. Many of these units often need personnel, computer equipment, and training.

- The forensic community recommends that the Federal Government require interoperability between Automated Fingerprint Identification System (AFIS) systems of different manufacturers to enable an "enter once, search many" capability. AFIS is a computerized biometric database system for electronically encoding, searching, and matching fingerprints associated with the investigation of a crime. Fingerprint units need computer equipment for systems such as AFIS and better networking and connectivity to State and regional AFIS systems. Roughly half of the fingerprint units have AFIS capability. Although technically feasible, no requirement exists for interoperability among AFIS systems of various manufacturers, so it is not possible to enter a fingerprint in one State and search that fingerprint in another State.

- The forensic community urges the Federal Bureau of Investigation (FBI) to continue to install Universal Latent Workstations (ULWs) in local agencies to maximize the investigatory potential of the Integrated Automated Fingerprint Identification System (IAFIS). It should be noted that the FBI, through the Criminal Justice Information System (CJIS), provides ULW connectivity through the CJIS Wide Area Network (CJIS–WAN), and continues to do so. It was reported by IAI that IAFIS is greatly underutilized for latent fingerprint identification. Most State agencies have access for latent searches through latent terminals, equipment which is beyond the reach of most local agencies. Local agencies must, for the most part, rely on State agencies to receive and enter their latent prints into the IAFIS latent terminal. Therefore, the tool designed for local agencies, the ULW, is underutilized for a variety of reasons such as lack of network access, lack of training and technical expertise, and lack of knowledge respecting how to solve those problems.

- In the area of footwear/tiretrack impression evidence, the forensic community urges funding for research, training, a Web-based footwear database, and a scientific working group. Footwear impression, barefoot impression, and tire impression evidence are much underused forms of evidence and are often overlooked. The forensic community believes that this under use results from the lack of an aggressive attitude about the detection and recovery of this evidence. Generally, there is little indepth training in the recovery and preservation of this evidence for crime scene technicians and police officers. In cases in which impressions are located and recovered from a crime scene, the quality of the recovery is often inferior, leading to unusable or less definitive evidence. Resources to recover this evidence (electrostatic lifters, lifting films, and casting materials) are often not available.

- The forensic community urges more personnel and better equipment and training for those who process and collect evidence from crime scenes. Most crime scene processing is done outside the crime laboratory by sworn law enforcement officers rather than specialized crime scene investigation units or evidence technicians. Equipment such as digital cameras, laser survey/mapping equipment for diagramming crime scenes, vehicles, and alternate light sources are tools not readily affordable by many agencies, but would ensure the integrity of evidence.

- Like many forensic disciplines, lack of manpower and equipment are concerns in digital forensic science. According to studies by NIJ and the Institute for Security Technology Studies (ISTS) at Dartmouth, the law enforcement community has identified a need for more computer crime investigators and technology/equipment. Lured by shorter hours and higher pay, highly trained officers often elect to enter the private sector, further accelerating the manpower shortage. Rapid change in computer-related technology also quickly leads to outdated equipment, technology, tools, and techniques. According to the ISTS studies (2002 and 2004), 41 percent of the respondents indicated that the current tools lacked essential features and 40 percent indicated that tools did not exist for functions they needed to carry out as part of their investigative process.

Medicolegal Death Investigation

Medicolegal death investigation is an essential justice and health function whose professionals play an important role in determining the cause and manner of death. According to NAME, the medicolegal death investigation system in this country is a frayed patchwork: 21 States have medical examiner systems, 11 have coroner systems, and 18 have mixed systems. Regardless of the type of system, only approximately half of the population of this country is served by systems with forensic pathologists. In most States, medicolegal death investigation is conducted by county offices that often cannot directly support complete death investigations. Although many medicolegal offices are of high quality, others lack funding, competent staff, and facilities. The forensic community strongly supports improving medicolegal offices to ensure appropriate death investigation throughout the United States.

Workload. As of 2003, the United States had 989 board-certified forensic pathologists. The forensic community reports that approximately 600 of those forensic pathologists are active practitioners, and less than 400 function as full-time dedicated forensic pathologists working within medicolegal death investigation systems. Besides limited availability of forensic pathologists, the forensic community believes that many current practitioners are exceeding recommended caseloads and many medicolegal autopsies are being conducted by nonforensic pathologist practitioners. This practice can result in increased errors, autopsies being performed by unqualified personnel (or not being performed at all), and manpower burnout and attrition.

Over the past 25 years, NAME has studied staffing requirements and workload capabilities for medicolegal offices and forensic pathologists. Based on these studies, NAME has recommended that a forensic pathologist who has no administrative duties should perform no more than 250 autopsies per year. When the number of autopsies performed exceeds this threshold, a forensic pathologist, no matter how skilled, may engage in shortcuts (e.g., performing partial autopsies when a full autopsy is warranted) or make mistakes. By the time the workload exceeds 350 autopsies per year, NAME believes that mistakes are more likely to be flagrant and involve errors in judgment (e.g., a case may not be autopsied that should have been or a diagnosis may be made hastily without sufficient basis, thought, or circumspection). Substandard work can result in faulty attributions of blame, wrongful prosecution or exonerations, and missed homicides. NAME believes that the United States should have a workforce of at least 850 full-time, board-certified forensic pathologists to maintain medicolegal autopsy loads.

Compensation. The most costly feature of upgrading and running a medicolegal office is the compensation of medical examiners. The average salary for a hospital pathologist in the United States is $270,000. However, in many areas of the country, chief medical examiners earn less than $150,000 and some medical examiners make $120,000 or less. NAME believes that this salary range is depressed when compared to other medical salaries; therefore, the number of medical examiners is not likely to increase.

Approximately 30 to 40 forensic pathologists are trained annually, but one-third of these practice hospital pathology only or forensic pathology only part-time. Another third of forensic pathologists drop out of practice within 10 years. Low salaries contribute to medical examiner offices traditionally drawing a small core of highly qualified dedicated individuals and a host of people with marginal qualifications.

Equipment and facilities. The forensic community believes that a medical examiner's office should consist of four minimum components: medical, investigative, administrative, and technical support. NAME also asserts that investigators should be trained properly in medicolegal investigation. NAME also urges that investigators should be employees of the medicolegal death investigation system, not law enforcement agents. NAME also suggests that the toxicology laboratory should be on the premises and under the authority of the chief medical examiner. NAME also notes that other supporting laboratory functions without particularly unique features in a medical examiner setting, such as histology or microbiology, may also be a part of the medical examiner office or those services may be obtained by contract.

Thirty-eight percent of the medical examiner offices surveyed by NAME did not have in-house toxicology laboratories, and some were thus dependent on State or police crime labs. These outside tabs can take several months to a year to report results, posing difficulties for case disposition. Moreover, the forensic community notes that crime labs perform limited toxicological analyses, using methods not sanctioned by the American Board of Forensic Toxicology (ABFT), which can lead to incomplete toxicological results. Other medical examiner offices must rely on private toxicology laboratories or clinical laboratories. NAME believes that it is desirable for all medical examiner offices to have dedicated support laboratories and appropriate toxicology professionals in house. The basic equipment cost to set up an in-house toxicology lab to handle 400 autopsies per year is more than $300,000. However, many jurisdictions cannot afford to equip or staff in-house toxicology laboratories. In addition, accreditation demands can also increase costs.

NAME notes that medicolegal offices often are poorly equipped and inadequately housed. Responses to a recent NAME survey of 128 medical examiner and autopsy-performing coroner offices revealed that 8 percent of them did not have the x-ray equipment necessary to make basic diagnoses or locate radio-opaque objects such as bullets. Significant numbers of forensic autopsies are done in funeral homes that lack necessary equipment such as x-ray equipment, adequate lighting, and scales to weigh the body and organs.

Poor facility design can have other adverse consequences. Heating, ventilation, and air conditioning (HVAC) problems allowed the spread of tuberculosis in more than one medical examiner's office. Facilities still exist that lack drains. In these cases, blood and other body fluids that can contain infectious material are sometimes collected in buckets and disposed of in sinks or toilets. At least one-third of facilities lack appropriate design and airflow systems to facilitate control of airborne and other pathogens.

NAME urges the appropriate distribution of forensic pathologists throughout the United States so they are readily available to all systems. A national strategy could include the use of regionally based investigation systems.

2. Continuing Education

The forensic community reports that training needs are significant and vary across disciplines. Providing training for novices and continuing education for seasoned professionals ensures that crime laboratories deliver the best possible service to the criminal justice system.

To be competent to analyze evidence, forensic examiners need both basic scientific education and discipline-specific training. To be in compliance with widely accepted accreditation standards, scientists in most of the disciplines must have, at a minimum, a baccalaureate degree in a natural science, forensic science, or a closely related field of study. Education and training are also needed to maintain expertise, update knowledge and skills, and keep up with advances and changes in technology.

When a new analyst or examiner is hired, that individual requires initial training to build competency. The length of the initial training also depends on the laboratory specialty area. For example, controlled substances analysts may require only 6 to 12 months of training. Thus training in experience-based disciplines such as latent print examinations, firearms and toolmarks analyses, and questioned documents examinations may require up to 3 years of training before being permitted to perform independent casework. Requirements for continuing professional development training may vary by forensic discipline.

Prior to conducting analysis on evidence, forensic scientists require both basic scientific education and discipline-specific training. To be in compliance with widely accepted laboratory accreditation standards, forensic scientists working in crime laboratories must have, at a minimum, a baccalaureate degree in a natural science, forensic science, or a closely related field of study. Each examiner must also have successfully completed a competency test (usually after a training period) prior to assuming independent casework. Education and training also are needed to maintain expertise, update knowledge and skills, and keep up with advances or changes in technology.

These needs can be addressed by collaborations, innovative approaches, and alternative delivery systems for forensic analysts and manager training. Regional centers based on established programs could also be used for expanded training. Professional models for training and establishing competency should be developed.

Forensic Science Education

Although the number of forensic science programs at colleges and universities has recently increased, the Council on Forensic Science Education (COFSE) has noted that many forensic educational programs have been established with very limited resources, insufficient personnel, laboratory space, and support. NIJ's Technical Working Group on Education (TWGED) has recommended guidelines for forensic science education programs. It provides minimum curricula guidelines for undergraduate and graduate science programs. TWGED also recommends that academic forensic science programs establish a working relationship with forensic science laboratories and that forensic science educational programs seek accreditation.

In 2002, the American Academy of Forensic Sciences (AAFS) established the Forensic Educational Programs Accreditation Commission (FEPAC) to establish a program for formal evaluation and recognition of college-level academic programs based on the TWGED guidelines. With financial assistance from AAFS and NIJ, FEPAC established standards, policies, and procedures to accredit university forensic science programs. The program includes a self-study completed by the university applying for accreditation and an onsite assessment by trained FEPAC assessors. In 2003, a pilot test of the FEPAC accreditation program resulted in the accreditation of forensic programs at five colleges/universities. Pilot testing of this program continues.

AAFS and NIJ provided financial assistance for pilot accreditations. As a result, costs for these accreditations are reduced during the pilot stage of this program. In order to ensure that our forensic scientists of tomorrow are adequately and uniformly equipped today, the forensic community urges continued support for FEPAC. This support will assist the community by keeping the costs of the program affordable or universities and colleges that seek recognition for their forensic science programs. Additionally, FEPAC is currently focused on university programs with traditional delivery systems. The forensic community believes that the program should be expanded to consider less traditional program delivery mechanisms, including distance learning.

The TWGED guidelines recommend that institutional support for forensic science programs be comparable to other natural science programs. Graduate education in forensic science has not received dedicated criminal justice funding, although educational loans and other forms of financing are well established for other graduate programs throughout the country. NIJ has traditionally supported graduate programs by providing research funding for the forensic sciences. A program to eliminate or forgive student loans for those graduates who obtain full-time employment in public forensic science institutions would be one such alternative source and should be considered. Any support provided would need to ensure that it is directed to those who would be employed in the public criminal justice sector.

In addition to research and student support, the forensic community seeks support for the acquisition and maintenance of equipment, for major research instrumentation, and for laboratory renovation. Institutions offering forensic science programs should address the ongoing costs associated with the important practical laboratory components of their programs. The typical cost for the research component for a master's degree thesis, a requirement to meet FEPAC accreditation standards, is between $15,000 and $20,000 per student, in addition to other tuition and educational costs each student will incur. In order to ensure the integrity of forensic science educational programs nationwide, the forensic community believes that any government resources that support university forensic science programs and students should be linked to FEPAC accreditation.

Forensic Science Training

To be in compliance with widely accepted accreditation standards, scientists in each of the disciplines must have, at a minimum, a baccalaureate degree in a natural science, forensic science, or a closely related field of study. However, to be competent to analyze evidence, forensic scientists need both basic scientific education and discipline-specific training. Hands-on training is needed to develop and maintain expertise, update knowledge and skills, and keep up with advances and changes in technology.

Initial training. When a new analyst or examiner is hired, that individual requires initial training to build competency and proficiency with standard operating procedures. The length of the initial training provided to an analyst depends on the discipline the trainee will enter, and operating procedures may vary from laboratory to laboratory within a specific discipline. For example, controlled substance analysts may require only 6 to 12 months of training. Those training in experience-based disciplines such as latent prints examinations, firearms and toolmarks analyses, and questioned-documents examinations may require up to 3 years of training before being released to perform independent casework. During their training period, individuals in experience-based disciplines serve much like an apprentice to a senior examiner.

Initial training remains largely on-the-job and is labor intensive. The laboratory manager must first identify an existing member of their staff with appropriate expertise and experience who can serve as the trainer. Often, this is an individual with significant casework experience whose casework productivity is reduced or lost to the laboratory during the training period. Laboratory accreditation standards require the training to be documented and to contain a demonstration of competency prior to assuming casework responsibilities. The salary cost of an analyst in a 1-year training program is between $30,000 and $40,000, but the cost to the laboratory is equally significant as laboratories can realize up to a 30-percent reduction in productivity during that training interval.

Some visiting-scientist and intern programs are available that can be used to augment or abbreviate initial onsite training, but costs are high and funding remains scarce. Some laboratories (e.g., the State laboratories in Illinois and Virginia) have begun collaborations with universities to offer their initial training programs to students enrolled in the university's graduate program. For example, through the residency program, qualified students at the University of Illinois in Chicago receive the same initial training provided to employees of the Illinois State Police Department, with the exception of supervised casework. In such a program, the agency providing the training does not pay a salary to the individual during the training, although Virginia does pay some stipends, lowering the cost of training considerably and greatly reducing the training burden on experienced examiners.

Some crime laboratories have made attempts to collaborate on initial training, sending the individuals to be trained to a single site. The Illinois State Police has accepted individuals from other States/laboratories into their training programs when space exists. The National Forensic Science Technology Center (NFSTC) has developed an academy program as part of its cooperative agreement with NIJ. NFSTC academies typically run for 16 weeks and provide intensive programs of study for new recruits to crime laboratories. Thus far, NFSTC has designed and presented Drug Chemistry and DNA Analysis Academies. An Academy in Forensic Firearms Examination is under development. After the pilot testing of an academy program, the NFSTC will no longer offer the training as part of its cooperative agreement. It will make the curricula available to the community for use in their laboratories.

Continuing professional training. Training also is required on a continuing basis for qualified analysts to maintain and update their knowledge and skills in new technology, equipment, and methods. Almost all scientific and technical working groups, certification programs (e.g., American Board of Criminalistics and International Association for Identification), and accreditation programs (e.g., the American Society of Crime Laboratory Directors/Laboratory Accreditation Board (ASCLD/LAB) and Forensic Quality Services) recommend or require continuing professional development training, but the requirements vary by discipline. TWGED provided an outline of criteria for continuing professional training to be used as a guide to provide a common framework to ensure that programs contain essential elements. ASCLD's effort to develop a model evaluation program for training includes continuing professional development programs. Currently, there is no funding source for such a program.

Symposia, workshops, and short courses are offered on a number of topics by an array of service providers to include professional societies and associations; NIJ's Forensic Resource Network (FRN); and Federal (e.g., FBI, Drug Enforcement Administration (DEA)) and State laboratories. Agencies often pay travel costs of $1,000 or more per person. It should be noted that training by federal agencies such as the FBI and the DEA is often delivered no cost to the agency.

Assistance has been provided to the crime laboratory community through a variety of programs, including FRN and grant programs from NIJ. These programs have been invaluable to the community, providing resources and training to address issues ranging from quality systems, training models, accreditation, and certification.

The cost of continuing professional development varies, depending on the requirements of the specialty. For example, the Scientific Working Group on Analysis of Seized Drugs (SWGDRUG) recommends a minimum of 20 contact hours per year for each analyst. The FBI's *Quality Assurance Standards for Forensic DNA Testing Laboratories* require a minimum of eight (8) hours of continuing education on an annual basis. The ASCLD/LAB accreditation program has adopted this latter requirement for DNA analysts.

The TWGED recommended that between 1 percent and 3 percent of the total forensic science laboratory budget be allocated for training and continuing professional development. Preliminary data reported by BJS from its crime laboratory census showed that the training and continuing education budgets of the largest 50 laboratories in the United States were actually less than one-half of 1 percent of their total budgets, In lieu of time requirements or a percentage, some agencies specify a budget amount for each analyst per year. Considering that the funds support travel and fees, $1,000 to $1,500 per analyst per year is typical. For a laboratory with 25 analysts, the annual cost of continuing professional development would be an estimated $25,000.

In addition to technical training (either initial or continuing), analysts need ongoing professional development training in a wide range of topics, including ethics, courtroom testimony, quality assurance, and safety. Some agencies (e.g., Illinois State Police and Virginia Division of Forensic Sciences) include this type of training as part of agency training programs. Professional organizations such as AAFS and some regional forensic science societies also offer training opportunities that may include presentations or workshops on these topics. Travel costs (estimated at $1,000 per person) comprise a large portion of the costs for these programs.

Supervisors and managers often are educated in the sciences, but the forensic community also urges instruction in basic business and personnel management, fiscal procedure, and project management. Annual management symposiums are held by the FBI and ASCLD. In 2002 and 2003, more than 350 managers and supervisors attended each symposium, demonstrating the overwhelming need for training for all levels of management within forensic organizations. The FBI covers both transportation and on-site costs for state and local agency attendees to its management symposium. Funding for the ASCLD symposium was provided by NFSTC in cooperation with NIJ. The cost to attend the ASCLD symposium was approximately $1,225 per person. Through the cooperative agreement, that cost was reduced by approximately $400 for attendees, who received a housing allowance provided by NFSTC. Training symposiums also are scheduled for 2004.

Alternate delivery systems for forensic science training, such as electronic media, are increasingly being used. In April 2003, the FBI announced the FBI Virtual Academy, offering Web-based access to training. Additionally, the FBI is attempting to establish training partners to work together to standardize key curricula, using TWGED documents as a guide. The NFSTC is developing and testing a CD-based Quality Documents program that is being recommended for use in the ASCLD Accreditation Mentoring Program. Distance learning also is being developed for forensic science training. For example, several States such as Illinois (the Illinois State Police) use video conferencing in conjunction with onsite facilitators to allow its training coordinators to deliver training to multiple sites simultaneously. In this way, the number of trainees may exceed the capacity of a single site or small numbers of trainees may receive a standardized training presentation.

Certain types of training, however, require face-to-face or hands-on participation and evaluation. For these types of training, regionally based programs would reduce travel costs. Illinois, Virginia, New York, Florida, and California have operational laboratories/systems with well-developed training programs that also have strong collaborations with universities. Such established programs are ideally suited for expansion to provide training on a regional basis, if sufficient funding is provided.

The FBI's traditional, on-site, forensic training classes have been highly regarded within the forensic community, and for many agencies these opportunities provide the only technical training available within their budget constraints. The training has been offered at both the FBI Academy at Quantico and through "road schools," where the training came to various parts of the country. These courses have been offered at no cost to the attendees, as the FBI covered airfare, lodging, and meals. As expectations grow within the judicial system and technology continues to advance, there will be an increasing demand for these types of training opportunities. The forensic community urges that funding be provided so that technical training can be expanded to meet the demand for on-site training. It should be noted that in fiscal year 2003, the FBI provided 1,311 law enforcement training opportunities of various types to non-FBI personnel. In fiscal year 2004, it provided 2,857 such opportunities, including 1,150 attendees of the Society of Forensic Toxicologists Conference which it co-sponsored with NIJ. In addition, FBI Laboratory personnel provided presentations to more than 5,000 attendees of meetings and more than 2,000 attendees of workshops or road show schools.

Investigators in the newest forensic discipline, digital evidence, also need to remain current in a fast-changing field. The discipline is now accredited by ASCLD/LAB, but currently there are no nationally recognized standards or certification for digital forensic practitioners. Some digital evidence investigation is done in crime laboratories, but most is conducted in law enforcement agencies similar to fingerprint units.

For digital evidence investigation, most law enforcement officers receive training in-house; through grassroots organizations (e.g., High Tech Crime Investigator Association, International Association of Computer Investigative Specialists, National White Collar Crime Center); vendor's software; and universities (e.g., University of New Haven, University of Central Florida). A minimal number of colleges, universities, and training facilities have curriculums devoted to computer forensics and other aspects of digital evidence. Continuing education programs exist but in limited numbers.

ASCLD/LAB accredited the first crime laboratory in digital evidence in December 2003 (North Carolina Bureau of Investigation, Raleigh). Other State and Federal laboratories are preparing for accreditation in digital evidence. The expense of digital equipment and facilities, combined with the level of ongoing training required to effectively stay current with advancing technology, hamper the development of digital evidence units in the majority of crime laboratories.

In bloodstain pattern analysis, demand is increasing for more trained personnel to assist in crime scene reconstruction. Bloodstain pattern evidence analysis is a forensic discipline generally performed by highly trained laboratory specialists employed by State crime laboratories or large law enforcement agencies with crime laboratory and crime scene capabilities. This type of evidence is often destroyed at the scene, poorly documented, or goes unrecognized as potential evidence due to lack of knowledge and training by the initial investigators and crime scene personnel. Bloodstain pattern analyses are dependent on good crime scene documentation, collection of relevant evidence such as blood swabs, "mapping" of bloodstain patterns with scales, 90-degree photography, and other ancillary documentation. Without proper bloodstain pattern training of crime scene personnel in the recognition and documentation of bloodstain patterns at crime scenes, even the most experienced bloodstain pattern practitioner cannot provide a useful analysis.

A change in emphasis has been made toward establishing standards, procedures, and educational requirements for the certified bloodstain pattern practitioners. The overall goal is to provide uniformity in bloodstain pattern analysis, to include terminology, training, education, casework examination, courtroom testimony, and research within the discipline. A defined basic entry level into this discipline is needed, with a process of continuing education and training until the bloodstain pattern examiner reaches the level of the certified practitioner. SWGSTAIN, an FBI sponsored scientific working group, has been actively pursuing these goals. Certification in bloodstain pattern evidence can be obtained through the IAI as a Certified Bloodstain Pattern Examiner.

Medicolegal Death Investigation Training

Forensic pathology is a recognized area of special competence within the field of pathology. It requires additional training and experience. Forensic pathologists must complete a standard pathology residency, complete an additional year of forensic pathology training, and pass examinations in both anatomic and forensic pathology in order to become board certified. Until recently, persons could also qualify to sit for the forensic pathology board examination by documenting sufficient experience within the field of forensic pathology, but this option is no longer allowed.

The American Board of Pathology (ABP) defines the educational and training requirements of this field and has provided specialty certification in this area since 1959. Most forensic pathologists undergo at least 9 years of formal education after college, including a medical degree, postgraduate residency in pathology, and additional formal training in forensic pathology and medicolegal death investigation, after which they must pass examinations in anatomic and forensic pathology in order to become board certified by the ABP.

As in other forensic disciplines, continuing education of forensic pathologists remains of great importance. Forensic pathologists not only require the services of the crime lab, but also are themselves forensic scientists who conduct their own forensic investigations. At the least, forensic pathologists need to be aware of the forensic laboratory analytic capabilities that can be applied to evidentiary material found on bodies, know how to conduct a thorough examination, and know how to collect, preserve, and document evidentiary material. This requires knowledge of current forensic science principles and capabilities. The forensic sciences have been greatly expanding and maturing in recent years, and it has been difficult for forensic pathologists to keep current with this burgeoning field.

Support for educational activities, national meetings, and research are all methods of fostering continuing education for forensic pathologists. Although NAME, AAFS, IAI, the American Society for Clinical Pathology (ASCP), and the College of American Pathologists (CAP) have forensic conferences and continuing education programs, all are cash-strapped. Also, continuing education costs run approximately $1,500 per year for each forensic pathologist, investigator, toxicologist, and administrator. Many offices do not defray or reimburse these costs, thus shifting the burden to individuals. Therefore, NAME recommends support for continuing education and encourages participation in professional meetings and conferences. Better forensic pathology education is a vital component of the goal to establish high-quality death investigation practices in every jurisdiction. NAME feels that the Federal Government can support the development of curricula following the Technical Working Group on Forensic Science Education (TWGED) model and establish Federal loan forgiveness programs for persons who become employed as government-paid medical examiners in geographic areas of critical need.

NAME also believes that the Federal Government should encourage competent death investigations by providing support to States that require and provide certified medicolegal death investigator training in accordance with the NIJ's *Death Investigation: A Guide for the Scene investigator*. It could also support forensic pathology education, experience for all anatomic pathology residents, forensic pathology fellowship training, and research.

3. Professionalism and Accreditation Standards

Maintaining and increasing professionalism within the forensic science community requires attention to a wide range of issues. Professionalism is supported by quality assurance measures such as laboratory accreditation and examiner/analyst certification, and the activities of professional organizations that establish scientific guides of best practice. Research, innovation, and technology transfer also are elements of professionalism practices.

Crime Laboratory Accreditation and Peer Certification

Crime laboratory accreditation requires laboratories to have and follow written policies to monitor quality. Accreditation requires a laboratory to evaluate its operations and, if problems are identified, address them. The largest accreditation program for crime laboratories in the United States is the ASCLD/LAB program. This program is currently in the process of establishing compliance with the International Organization for Standards (ISO). The NFSTC offers an ISO-compliant program for accrediting forensic laboratories.

Presently, 260 crime laboratories are accredited by ASCLD/LAB. At least three States have mandated the accreditation of their crime laboratories: New York, Texas, and Oklahoma. Nine States, however, do not have accredited laboratories.

The laboratories accredited by ASCLD/LAB are considered to be traditional crime laboratories. A traditional crime laboratory is a single laboratory or system comprised of scientists analyzing evidence in at least two of the following disciplines: controlled substances, trace, biology, toxicology, latent prints, questioned documents, firearms/toolmarks, or crime scene. It should be noted that not all forensic services are performed in what is thought of as a traditional crime laboratory. These forensic services may be provided through a site or unit comprised of sworn law enforcement personnel who may not have scientific training. Analyses in the disciplines of digital evidence, latent prints, questioned documents, and crime scene are usually conducted in identification units often found outside of the traditional laboratory setting, such as within a police agency.

If the definition of a crime laboratory is expanded to include identification units operating in the 14,000 police departments and law enforcement agencies in the United States, there could be as many as 1,000 forensic service providers. The actual total is unknown. The average size of traditional laboratories is 30 personnel (25 of whom would be considered analysts). The average size of the non-traditional crime laboratory is estimated to be three.

Increasing emphasis is being placed on accreditation and meeting quality assurance standards for crime laboratory operations, but many laboratories lack the funding to carry the process to completion. Many laboratories now face stagnant budgets and rising caseloads. Nevertheless, accreditation is viewed as an important credential for crime laboratories. The forensic community urges support for training and preaccreditation assessments.

ASCLD has established a formal mentoring program to assist its members in achieving accreditation by pairing a nonaccredited laboratory director with an accredited laboratory director. Participants in this program report the greatest impediments to accreditation are related to resources: both the personnel needed to work on the accreditation standards and the cost of the program itself.

Fees and inspection expenses are associated with a laboratory's participation in an accreditation program. An inspection team audits the laboratory only after submission of all required paperwork and following consultation between the laboratory and lead assessor. The size of the laboratory and the number of disciplines in the laboratory dictate the size of the inspection team and the time on site.

Certification within the forensic community is a voluntary process of peer review by which an individual practitioner is recognized as having attained the professional qualifications necessary to practice in one or more disciplines. The forensic community supports certification, but certification has associated costs. In addition to the initial cost of application and testing, the cost for the academic degree and continuing education necessary for certification are substantial when considering a large number of examiners. These expenses can have a significant effect on laboratory budgets that are currently struggling to meet the primary demands of casework.

There are a number of forensic certifying boards used to identify whether practitioners meet certain standards. The Forensic Specialties Accreditation Board (FSAB), created through AAFS, has worked to develop standards and a voluntary program to assess, recognize, and monitor the forensic specialty certifying boards. This process relies on international standards (ISO) and standards from other recognized accreditation bodies.

Forensic certifying boards are invited to participate in the FSAB if they meet established requirements including periodic recertification, an examination covering the knowledge base of the relevant forensic specialty, a process for providing credentials, and a code of ethics. Nine organizations representing the majority of the recognized boards offering forensic specialty certification were invited to join FSAB. These include the American Board of Criminalistics (ABC), the American Board of Forensic Document Examiners (ABFDE), the American Board of Forensic Odontology (ABFO), the American Board of Forensic Toxicology (ABFT), the American Board of Medicolegal Death Investigators (ABMDI), the Association of Forensic Document Examiners (AFDE), the Forensic Toxicologist Certification Board (FTCB), the International Association for Identification (IAI), and the International Institute for Engineering Sciences (IEES).

In a 2003 survey of 229 certification programs, the American National Standards Institute (ANSI) found that certification programs have a significant impact on a profession *(Personnel Certification: An Industry Scan,* 2003). That same study found the benefits of certification include enhanced credibility of certificants. Benefits also include the enhancement of professional development and training, as well as enhancement of academic training for the profession. Certification programs are often expensive to develop and administer. They also require substantial time and financial commitment by individuals participating in the programs.

The forensic community seeks support for forensic certification programs and the FSAB. Start-up costs for developing a certification program are considerable. Most programs are started with the expectation that they will become self-sustaining within 5 years of the first test administration. However, only about half of the organizations report achieving self-sustaining status within that period.

Medicolegal Accreditation

NAME has a broad-based inspection and accreditation system for medical examiner systems. This system examines facilities, safety, personnel, death notification, case acceptance, release of human remains, investigations, evidence and specimen collection, support services, reports and records, mass-disaster planning, and quality assurance. However, the majority of medical examiner offices in this country have not attained NAME accreditation. Often, accreditation is not feasible because of inadequate staffing, inadequate facilities, inadequate equipment, or a combination of these factors. Only 40 offices in the United States are accredited out of a total of 465 facilities. Many autopsies are performed in areas remote from accredited medicolegal facilities. Only 23 percent of the population is served by an accredited facility.

The accreditation process is difficult, time consuming, and costly. The forensic community reports that some offices obtain increased political and financial support as a result of the accreditation process. The community also reports, however, that there are few tangible incentives for accreditation other than assuring the community that the office uses best professional practices.

Many States have created statutory requirements for death scene investigators. For example, Tennessee requires 100 hours of certified training, and Indiana requires completion of a 40-hour training program supplemented with standardized testing. Many other State medical examiner and coroner associations and academic institutions have provided various levels of training for death investigators.

More and more jurisdictions are choosing to use "lay" (nonphysician) investigators to perform scene and background investigations in support of physician medical examiners and forensic pathologists. The American Board of Medicolegal Death Investigators (ABMDI) registers and certifies such practitioners in accordance with the National Institute of Justice's *Death Investigation: A Guide for the Scene Investigator*. This, however, is a voluntary program, and in some jurisdictions, investigators are not required to have any formal education in basic death investigation procedures.

One of the largest challenges to appropriate death scene investigations is the shortage of qualified personnel and the funds to train them. NAME believes that death investigators at every level should have appropriate training and perform their duties in accord with professionally accepted standards. NAME urges Federal support for training and professional certification of death investigators.

Research, Innovation, and Technology Transfer

Traditionally, basic scientific research is performed at universities. Forensic science, however, is a very specialized applied science. Academic and forensic practitioner partnerships can bring the skills and strengths of both basic and applied science to a research program. Such partnerships exist within the forensic community where a strong forensic laboratory works closely with a well-established, graduate-level university forensic program.

The Forensic Resource Network (FRN), an NIJ program, is one example of a partnership among research institutes, technology centers, and crime laboratories that promotes the implementation of new technologies and model training programs for the forensic laboratory community. Its mission serves to bridge the gap between the forensic research community and operational crime laboratories.

The FRN consists of the Marshall University Forensic Science Center at Huntington, West Virginia; the National Center for Forensic Science at Orlando, Florida; the National Forensic Science Technology Center at Largo, Florida; and the West Virginia Forensic Science Initiative at Morgantown, West Virginia. Network members provide training for laboratory personnel, technology transfer services, methods research and development, methods testing and evaluation services, and analytical services for laboratory casework outsourcing. ASCLD serves as advisor to the FRN, providing feedback and guidance for their project proposals.

To keep up with changing technology, the forensic community urges the continued development of scientific guides of best practice. The development of these guides for best practice has been primarily accomplished in the United States by scientific or technical working groups. These working groups are composed of multidisciplinary professionals but also include forensic scientists with discipline-specific expertise. Most are supported by a Federal agency and operate under the constraints of that agency. These working groups usually meet at least annually to consider technical and quality-related issues.

Scientific and technical working groups serve a valuable role in the forensic community. They work to develop analytical guidelines, training and educational recommendations, and quality assurance guidelines. The recommendations of these groups can be expected to have a significant impact on other certification and accreditation standards, as well as on the introduction and acceptance of expert evidence in the criminal courts. A strong vetting process is needed for the guidelines developed by these working groups to make them truly representative of the entire forensic community and practical in a variety of jurisdictions with differing legal and policy constraints.

4. Collaboration Among Federal, State, and Local Forensic Service Providers

Federal crime laboratories collaborate with State and local forensic service providers in a variety of ways. Federal laboratories provide resources for research, training, and technology transfer.

Examples of Federal crime laboratory support include the following:

- Training for examiners of all disciplines as well as laboratory managers.

- Participation in research partnerships and validation projects with State and local laboratories.

- Maintenance and support of databases for linking firearms evidence (National Integrated Ballistics Integrated Network), fingerprints (IAFIS), and DNA (Combined DNA Index System).

- Establishment of regional mitochondrial DNA laboratories.

Federal laboratory support to State and local crime labs is necessarily constrained by budgets, staff expertise and availability, and space availability. The forensic community urges the creation of a formal mechanism, such as an advisory board or focus group, to facilitate coordination and communication between Federal laboratories and the forensic community concerning the needs and priorities of the State and local laboratories.

State and local crime laboratories conduct some testing for Federal agencies. The FBI Laboratory accepts casework from State and local crime laboratories in certain circumstances, for example, when the State or local laboratory does not have the analytical capabilities to conduct the required analysis. The FBI Laboratory has perhaps the largest forensic research budget of any agency.

Federal medical examiner support. The only current Federal medical examiner system is the Armed Forces Medical Examiner System (AFMES), a specialty operation that primarily serves military combat and training casualties. The forensic community has expressed interest in the establishment of a national support system. The community noted that one option for such a system would be the consolidation of the Disaster Mortuary Operations Response Team (DMORT) division of the National Disaster Medical System (NDMS), which could be mobilized to assist State and local agencies in times of a declared disaster.

The AFMES is a specialty operation that primarily serves military combat and training casualties in addition to traditional civilian forensic pathology. As the only Federal medical examiner office, the AFMES is often consulted by other Federal agencies and participates in many interesting and important cases. For example, the FBI has no internal medical examiner capability and will often consult with the AFMES for forensic pathology expertise. The AFMES also serves the pathology community through consultation, education, and research. The AFMES will review civilian cases submitted to them for a fee; however, the number of civilian cases submitted are relatively few. The most important area in which the AFMES has assisted local medical examiners is in mass disaster situation, again when requested and usually for a fee, where they do have considerable expertise.

DMORT can be mobilized to assist State and local efforts in times of a declared disaster and has provided a valuable service to jurisdictions in need. Its services are particularly valuable when a multiple fatality incident occurs in a coroner jurisdiction with no forensic pathology resources, training, or experience; but even well-resourced offices often have limited contingency capacity.

5. Summary of Forensic Community Recommendations

State and local crime laboratories are an integral part of the criminal justice system. The demand for crime laboratory analyses has increased, but State and local support has not always kept pace with this increasing demand. Crime laboratory backlogs cause significant delays in evidence being analyzed, resulting in investigation and court proceeding delays.

Although each of the forensic disciplines has specific personnel, training, equipment, and facility requirements, they share common concerns about adequate manpower, continuing education, equipment, and training.

Following are general recommendations suggested by the forensic community:

Manpower and Equipment Needs

- An organized attempt should be made to determine the quantity of forensic service providers outside crime laboratories.

- There should be outreach to all forensic service providers, including noncrime lab providers, to advise them of professional and governmental assistance programs.

- The needs of the forensic community should be monitored on an ongoing, systematic basis.

- Government support of AFIS systems should be contingent on interoperability between AFIS systems of different manufacturers, allowing an "enter once, search many" capability. This interoperability must address not only a seamless exchange of fingerprint data among States and among State and local systems; that same seamless interoperability must be developed among all State and local systems and the latent print search capability of the FBI's IAFIS system.

- A quality medicolegal death investigation system should be encouraged. Professional death investigation systems should examine the need for fully trained and qualified forensic pathologists with competent investigative and support staffs. Specifically, States should reexamine their current medicolegal death investigation systems to determine whether they can conduct appropriate, timely, and reliable death investigations.

Continuing Education

- Professional models for training and establishing competency should be encouraged.

- Collaborations, innovative approaches, recognized training centers, and alternative delivery systems for forensic analyst and manager training should be considered to reduce training costs.

- Quality graduate education in forensic science programs should be encouraged. A program to eliminate or forgive student loans for graduates who obtain full-time employment in public forensic science institutions is one alternative that should be considered.

Professionalism and Accreditation Standards

- The forensic community supports accreditation of organizations and certification of practitioners.

Collaboration Among Federal, State, and Local Forensic Service Providers

- A formal mechanism, such as an advisory board or focus group, should be established to facilitate coordination and collaboration between Federal laboratories and the forensic community.

- The forensic science organizations support the creation of a National Forensic Science Commission to assess the needs of the forensic science community and to stimulate public awareness of and interest in the uses of forensic technology to solve crimes. The commission should be tasked to undertake a comprehensive review of the role of forensic science in the criminal justice system, cost/benefit analysis of the value of forensic science to the administration of justice, needs of forensic science providers, and policy issues with respect to forensic science.

- Information sharing and coordination with Federal agencies should be supported.

Conclusion

Forensic evidence is the most important investigative tool available to our adversarial system of justice that can help identify the guilty and exonerate the innocent. Over the past 10 years, some forensic technologies have advanced far more rapidly than others. As a result, legal issues such as admissibility and practical issues such as technology transfer may be at the forefront within some forensic disciplines, while others have met these challenges and continue to build on successes. Further, the important role that the forensic sciences can play in investigating mass casualties and domestic terrorism adds a new dimension of application and coordination which must be considered in the broader context of issues that affect the utility of forensic evidence.

To address these and other critical issues, the President's DNA Initiative directs the Attorney General to form a National Forensic Science Commission, which will play an important role in the Department of Justice's goal of identifying critical issues facing our Nation's forensic services. The Forensic Science Commission will be designed to help policymakers assess the needs of the forensic science community and stimulate public awareness of the uses of forensic technology to Investigate, solve, and prevent crimes.

The Forensic Science Commission will bring together providers and end users of forensic services to address issues raised in this report (and others) that affect the quality and timeliness of our Nation's forensic services. The Commission will consider issues that affect all disciplines of the forensic sciences and make recommendations to improve public safety through maximizing the use of forensic evidence. Its multidisciplinary membership will facilitate the development of strategic partnerships that represent diverse opinions and perspectives, including those of law enforcement, practitioners, academicians, attorneys, judges and ethicists. Such partnerships will be able to discuss in a public forum complex issues that affect the furtherance and advancement of forensic practice.

NIJ recognizes that a number of social, ethical, legal, and policy issues may arise as a result of effectively and efficiently implementing recommendations for enhancing our Nation's forensic service providers. However, NIJ is confident that the Forensic Commission will serve to provide a cognitive process by which criminal justice professionals and the public can openly and fully deliberate these sensitive matters.

References

Bureau of Justice Statistics, *"50 Largest Crime Labs,"* 2002

Bureau of Justice Statistics, *"Census of Publicly Funded Forensic Crime Labs,"* yet to be published

National Institute of Justice, *"Electronic Crime Needs Assessment for State and Local Law Enforcement,"* 2001

National Institute of Justice, *"Forensic Sciences: Review and Status of Needs,"* 1999

Technical Working Group on Education and Training, Final Report

Forensic Educational Programs Accreditation Commission, Accreditation Documents

FBI, "Quality Assurance Standards for Forensic DNA Testing Laboratories"

American Board of Pathology, Certification Documents

American Society of Crime Laboratory Directors/Laboratory, Accreditation Board Accreditation Documents

ISO, Accreditation Documents

National Association of Medical Examiners, Accreditation Documents

Forensic Specialties Accreditation Board, Accreditation Documents

American National Standards Institute, *"Personnel Certificatio:; American Industry Scan,"* 2003

National Institute of Justice, *"Death Investigation: A Guide for the Scene Investigator"*

Disaster Mortuary Operations Response Team, Operational Documents

Council on Forensic Science Education, Operational Documents

www.ingramcontent.com/pod-product-compliance
Lightning Source LLC
Chambersburg PA
CBHW081810170526
45167CB00008B/3392